Positively Skewed

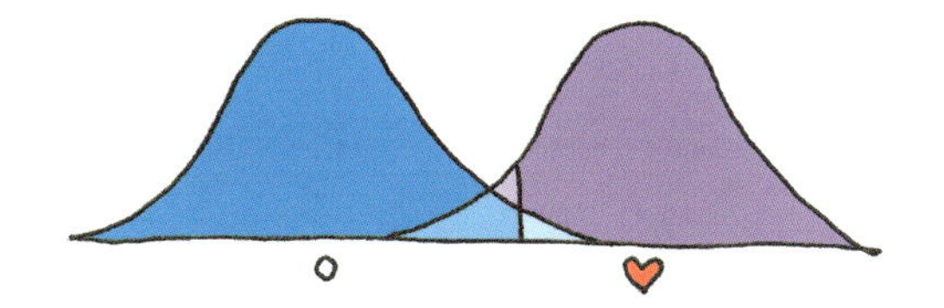

JULIET AIKEN

ISBN: 098891980X
ISBN-13: 978-0-9889198-0-8

For the Tomlinson Family

Once, there was a little sample.

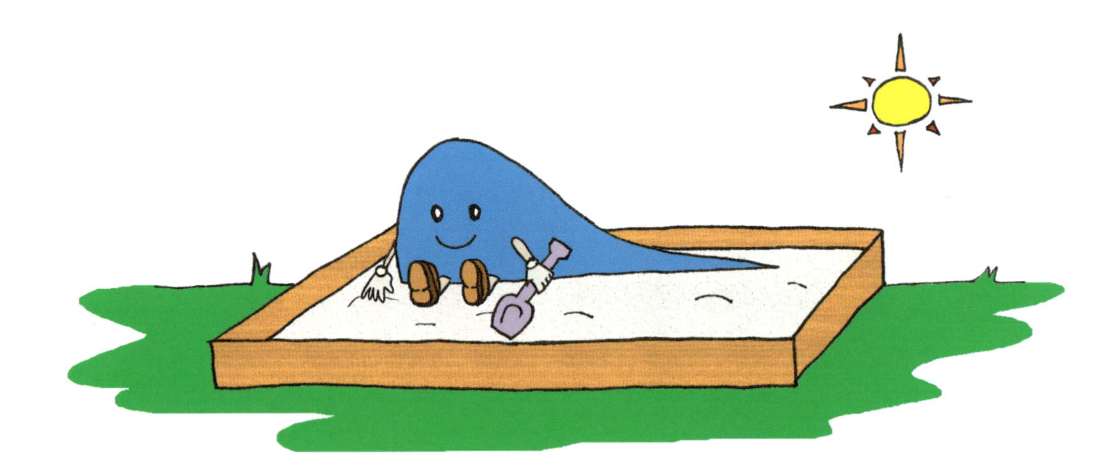

This little sample loved to be outside,

draw fun pictures on the wall,

and play with other little samples,

just like all of its friends.

But this little sample often felt lonely

because it just didn't look like its friends.

Its friends often joked that the little sample was skewed.

Not normal.

So the little sample stopped spending as much time with its friends.

It just didn't feel like it fit in.

One day, the little sample's population noticed that it was not spending as much time with its friends as it used to.

"What's wrong, little sample?" asked the population.

"I'm skewed," said the little sample.

"All of my friends tell me so."

"And what's wrong with that?" asked the population.

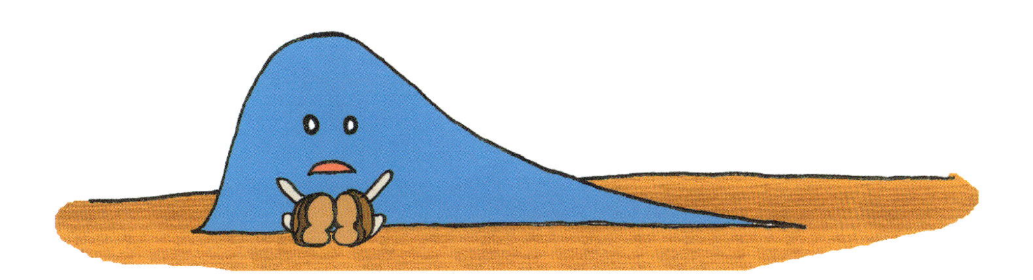

"You don't understand."

"I'm not normal like everyone else," said the little sample.

The population gave the little sample a biiiiig hug.

"Little sample," said the population,

"that is a good thing!"

"All samples are non-normal in their own way,"

"even your little sample friends."

The little sample could not believe that its friends were non-normal.

The next time it went outside, it looked closely at its friends . . .

and noticed that the population was right!

Some friends were bi-modal . . .

other friends were platykurtic . . .

and still others were slightly skewed . . . almost like the little sample itself!

The little sample never felt bad about being non-normal again

because every little sample is non-normal,

and that's what makes them so special!

18352809R00019

Made in the USA
Charleston, SC
29 March 2013